NATURE
ISN'T RACIST

BY: HARVEY BRUCE GRAHAM

ISBN: 978-1-965146-54-5

Dedication

I dedicate the contents of this book to the force that shapes our very form and whose laws govern our fragile existence: NATURE.

PREFACE

Too many times, man's evolutionary will to survive is being mistaken for racism.

About The Author

Harvey Bruce Graham grew up on a primitive farm in the bush on top of Blue Mountain in Ontario, Canada. His ancestors had migrated to this area in the late 1700s from Ireland and settled there.

This gave him ample time to study nature and be part of it. He learned that everything in nature has a part to play, and its most important job is survival until the next day.

06/09/2022 16:05

TABLE OF CONTENTS

CHAPTER ONE

Ever since life first occurred on earth, one activity has not changed: life forms that are similar in appearance tend to congregate together. This was done for a very good reason. Survival.

They have learned that life forms that appear different from them could harm their existence. Over time, the individuals that stayed with similar creatures survived to carry on the species, while those that didn't ceased to exist. That is why, to this day, you see robins with robins, rainbow trout with rainbow trout, and foxes with foxes. As a matter of fact, the only species that doesn't follow this lifestyle is humans. Why do we insist on being different?

We started on the right foot, and even in the Bible, they speak of the land of the Palestinians and the land of the Israelites, etc. Everyone stayed with their own kind. That is why there are so many countries in the world today. If we go to Germany, we find Germans. If we go to China, we see Chinese. If we go to Ireland, we find Irish, and so on. It's evident that we intended to carry on how nature intended us to, so what happened?

-2-

When the Europeans first discovered North America, they found that an indigenous race of tan-skinned humans already populated it. As their very survival depended on it, these people lived very close to nature and so they developed most of nature's habits. As a result, they were very cautious around these new humans, and once again, what nature had taught them proved to be true.

As it turned out, history has shown us that breaking from their policy of following the program laid out for all living things led to their troubled existence.

CHAPTER TWO

When the explorers returned to their home countries, word of the rich land they had discovered quickly spread to the West. This started a migration of settlers to the new world, looking to start a new life in this utopia across the sea.

As more and more immigrants came to North America and started to explore further and further west, they discovered that the indigenous population was not concentrated in one area but was spread out in groups across the country. These small groups had a leader and a name and were identified as bands or tribes. They had learned from observing nature that dense populations of any species only led to serious problems. They had observed that foxes, coyotes, and wolves were all canines, but they didn't live together. As a matter of fact, they couldn't. The competition for the food that the land could supply was too much, so each species had its own territory. It even went further than that. Each family group of each species had its own area.

The natives had no internet, television, or university to teach them. All they could do was learn from observing nature, and learn they did.

CHAPTER THREE

Although the natives were well aware of the dangers of having too dense a population, given the environment they were living in, this would have been almost impossible. Given the harsh conditions, no technology, no medicine, no tools, and only nature to teach them, a forty-year-old native was a very old and lucky person. Returning to the fact that they only had self-made remedies or medicines brings up another very good reason for all species to stick to their own kind.

The new invaders were spreading old-world disease among the natives at an alarming rate. Having no immunity, this was a disaster waiting to happen, and happen it did. Smallpox and other epidemics wreaked havoc on the unsuspecting population. Only the fact that they were so widely distributed throughout the country saved them from being totally wiped out. Even though they had broken nature's rule to avoid other species different from them, they had survived by heeding another rule of nature: to avoid dense populations.

No matter what form of life you are, the first thing you will see when you are born is something that will be very similar to you and that image is imprinted in your memory banks for life.

-2-

That is the number one reason that all species of every form of life on earth stick to the creatures they are familiar with for a sense of security. They had learned from the very beginning when the first primitive organisms crawled out of a primeval soup of amino acids that if you were not careful with what you associated with, it could very well lead to the end of your existence. The DNA formed from these encounters has been passed down to this day and regulates why all living things congregate with their own kind.

CHAPTER FOUR

As more and more explorers started to push into other areas of our planet, they discovered a very familiar behavior. When the Spanish began to drive deeper into South America, they discovered it was already populated by a dark-skinned race of humans that were all very similar in appearance. Still, they didn't all exist in one part of the country. Instead, they were widely dispersed throughout the continent, living in isolated groups or tribes. Even though these people had experienced no previous contact with the outside world, they instinctively knew that living together in a densely populated area could prove to be very problematic to their existence. From the Arctic to Africa, this behavior proved to be consistent.

Even though this knowledge is stored in our DNA, why do we continue congregating until we live in vast cities? The reason is quite obvious. When people come to a new land, they tend to settle in an area that is suitable to their needs – good soil, water, building materials, and easy accessibility. As more and more people arrive, they see that the area is already settled, so it must be a good place to stay. So they settle down there as well. As the area's population increases, the people need services, so businesses start up, roads are built, schools are established, etc. There you have it – the start of a city.

CHAPTER FIVE

Unfortunately, not all land on the earth's surface is habitable. Taking into account that 70% of the earth's surface is water and that 60% of the land surface is too harsh to live on, only about 15% of the planet is capable of human existence.

Given that the earth's population is growing alarmingly, we are heading to a definite problem. So-called racism is a symptom, not a disease. We started out right with similar races living in their own countries. Still, they have been forced to move to different countries with unfamiliar religions, customs, and lifestyles due to overpopulation, so it begins.

Everything in nature was designed to live with its own kind. The same basic laws apply, whether animal, insect, bird, reptile or human. Every species has its own basic needs and requirements from diet, climate, and living conditions that might not be acceptable to other species. You won't find marsh marigolds growing in the desert or palm trees growing in the Arctic to make the point that everything has its own needs simply. It is indeed the same for every race of human living on the planet. Once again it comes back to the fact that humans forget that we are a part of nature and we don't, and can't, control it, and we must follow its laws the same as every living thing does, or face the consequences. The fact that people are leaving their own countries to live in strange lands because of overpopulation proves that we have violated another of nature's laws. No species should overpopulate to the point that it creates a problem.

CHAPTER SIX

Every living thing except some lower life forms has a specific time of the year for mating for the purpose of carrying on the species. Not so with humans. Ever since we learned that sex was pleasurable, we engaged in it constantly, whether our population situation was critical or not. In fact, many species go one step further, and the males and females live separately except for the time of the year designated for breeding. This behavior enforces two laws of nature: keeping groups small to make existence less difficult and solving the overpopulation problem. The argument can be made that sex is so popular because it feels so good. Of course, it does. Nature intended it to be. If having sex felt like poking yourself in the eye with a sharp stick, we would not have to worry about overpopulation. We would all be extinct.

But that is not the case, and overpopulation is a major problem and one of the contributors to hate and racism crimes in the world today. Even though this is a fact and should be one of the major concerns in the world today, very little is ever mentioned about controlling the earth's population. Why?

CHAPTER SEVEN

As the human population increases and more and more of the natural planet is eliminated, there is a tendency to move away from existing with nature and simply eradicate it to make room for our exploding numbers. We are moving further and further away from co-existing with nature and following its rules and ways to rewrite them to suit our convenience. We will realize too late that this is a horrible mistake.

Nature loves a balance and everything in the universe strives to obtain this goal. Whether it is weather systems or animal populations, they are all seeking the same outcome. Weather does it by balancing barometric pressure; animals use instinctive behavior and the actions of other species to achieve the balance in their population that they need to survive.

Everything in nature has a saturation point, which is simply the time that a certain species can no longer survive in its present environment without making changes. Often, it is the ability of the area where the species resides to supply enough food to sustain it. This could be a variety of reasons, climate change or over-population. Moving to another area is a temporary solution for both situations, but it's not solving the problem. The changing climate is a difficult fix as over-population is a major contributor to the situation. There is a solution to the problem of having too many of any species, and it's not moving to another place. We are experiencing that, creating a new and dangerous situation. We've even created a word for it.

-2-

We call it racism. As I mentioned earlier, nature loves balance, which is very evident in its way of controlling its population. Everything is classified as either predator or prey, and their numbers are directly related to the ongoing cycle of life. For example, let's say the local fox population is in decline because of over-hunting by humans. This will trigger a natural response from the fox. First, the female will start to have larger litters; second, when the young foxes are reproducing, they will immediately leave their mom. They will go out and begin forming families of their own. These changes will start an upward swing in the local fox numbers. The increase in the predator density obviously has a negative effect on the prey community, and they, in turn, make changes to avoid being wiped out.

So on it goes, a perpetual cycle of highs and lows so that every living thing can survive and live on for another day. This is how nature maintains a balance. So, where do humans fit in?

CHAPTER EIGHT

"Birds of a feather flock together" is an often-heard old phrase that indeed can't be disputed. That is pretty much the way we started out, the way nature intended it to be. Unfortunately, except for natural disasters, old age, wars, and disease, we pretty much had no control over our population. Because there was so much space, people were encouraged to have large families to develop it how things have changed!

With all the advances and medicine, we have conquered many of the deadly diseases and people are living longer than ever. With the advances in science, we are being forewarned of many natural disasters and prepare for them in advance. The one thing we haven't learned to control is our birth rate.

All living things in nature face the same survival problems as humans, but they have one feature that we don't have. In difficult and unfavorable times, their bodies can regulate the number of offspring they produce. One example of this is the arctic ground squirrel. When times are tough, and the environment cannot support any more of their numbers, the females basically shut down reproduction, thus controlling their numbers. Various species use similar techniques to achieve the same goal. As for humans, controlling our population is the furthest thing from our minds. If you have another child, the government rewards you with another monthly check in your bank account. With incentives like this, it's hard to think about birth control.

-2-

In such a vast country as Canada, it's hard to visualize that we would ever have a problem with overpopulation. But here's the crux of the situation. Many countries have already reached saturation and are coming here in groves. Our vast land is suddenly shrinking extremely rapidly. It's a recipe for disaster.

Too dense a population, especially one made up of people who are different from each other, sets up a chain reaction of events, none of which is good. All of a sudden, the land that was being used to grow food was now needed to build houses. Basic needs are in short supply. We desperately need more food, fresh water, clean air, power plants, etc. Once the ball is rolling, there is no stopping it. Unless nature hits us with a disaster of some sort (for example, the pandemic) that wipes out half of our population, we need to save ourselves desperately. The planet isn't getting any larger.

CHAPTER NINE

With the influx of immigrants presenting a cross-section of all the countries on earth, Canada or any other country accepting them will be receiving a mixed bag of religions, customs, languages, and cultures that will all try to co-exist in a new country. At the same time, try to learn the ways of a new environment.

Our government tells us that it wants to build a diverse country where all the different races can live and work together in harmony. This would never work in nature and won't work for humans.

Take, for example, the birds in your backyard. There are probably several different varieties, but they will all mingle with their own kind. This will be especially noticeable if you have a bird feeder. Each species of bird prefers to eat with its own kind. They do not like to interact with the other species. This is not just a characteristic restricted to birds. It is a fact of all species of living things.

For example, let's take the fish species. There are thousands of different types of fish, so let's pick the trout. This fish species is broken down into many different breeds, but we will pick the brook trout. They inhabit the same water as other trout but have no interaction with the other species.

-2-

The brook trout and another fish, the lake trout, have qualities that fishermen demand, so biologists decided to do an experiment. They crossed the eggs from a lake trout with the sperm from a brook trout to create a new species. They called it the splake. The idea was to stock it in lakes and rivers and have a new fish for angling pleasure. Well, guess what! It didn't work. The new species they had produced was sterile and couldn't reproduce. Nature said, "You're not going to start messing with my creations."

It seems that all living things will go to great lengths to remain with others of their own kind. It is a very simple and efficient method when everyone is on the same page. A group of similar individuals will claim an area as their territory. This can be and is done by individuals of a species. Humans started out this way, but they marked their territory with borders, which is still done today. There is one thing that all people who leave their native land to settle in a new one have in common. When they arrive in their new homeland, they all congregate together and claim an area as their own. That is because of their inherent need to be surrounded by living things that are familiar to them for their own security.

Flashback to the caveman who got eaten by a dinosaur. The caveman who was watching this turned to his buddy and said, "Remind me to stay away from those things!" The situation I just described might not have played out in exactly that manner, but without a doubt, the message came across loud and clear.

-3-

Things that are different can be deadly. This image has been transmitted to our brains thousands of times and, over millennia, has become embedded in our DNA.

CHAPTER TEN

I am sure we all have experienced an uneasy feeling when starting a new job or moving to a new location. This is our body switching into survival mode. We are out of our territory, and everything can threaten our existence. We will continue to be on high alert until we have established a new area for ourselves and become familiar with the things and humans in it. This trait was passed down from our ancient ancestors, who had to constantly be in a state of survival mode if they wanted to survive.

This is another factor in the reason that when new immigrants show up in our space who may not look or dress the same as us, we will feel anxious and go into a state of extreme caution. It has absolutely nothing to do with hatred. It is simply a reaction to help increase the chances of self-preservation.

No matter what country you presently reside in, if your ancestors helped to create that country, carve it out of its rough form, and fight for its existence, there is no doubt that you feel strongly that this is YOUR country and will fight to keep it.

Nothing speaks stronger than the outcome of past actions. Nine hundred years ago, the very first settlers were arriving in the Americas from the old world and we all know what happened next. The indigenous natives of what is now the United States and Canada, being very close to nature, fought hard to retain their territory. In the end, they were outnumbered and resigned to live on reservations selected for them.

-2-

In Central and South America, this scenario was played out once again. Following tails of vast amounts of gold, the Spanish pushed deep into the jungles and discovered ancient tribes living on established territories as they had for hundreds of years. From then on, their way of life would never be the same. Were any of these acts either in South or North America driven by racism? Absolutely not. Neither the Spanish nor the settlers of North America had a clue who these people were, let alone hate them. They were simply an obstacle to overcome. So, what have all these past actions shown the people of today? Back to the caveman and the dinosaur, be very careful when confronted with something different.

There is another example of species invasion taking place in nature right before our eyes, and it involves every type of life form. We've even given them a scientific name. We call them invasive species. This organism takes up residence in a country it is not native to and wreaks havoc on the local environment. If it is a plant form, it establishes itself where the local vegetation grows and expands very rapidly, choking out the native plants. If it is a fish, it targets the marine life's food supply and eats the eggs of the local fish. The outcome is the same whether it is a bird, animal, or insect. Nothing good ever results from the arrival of a foreign competitor.

-3-

Whether we like it or not, we have all been an invasive species. Just ask the indigenous tribes in North America when the explorers first arrived to settle on their land. Once again, nature tried to warn them, and their survival instinct kicked in, but we all know the outcome.

CHAPTER ELEVEN

Every living thing has a survival instinct, and that includes all humans. That being said, it is evident that the immigrants coming to your country have the same misgivings as us as we do about them. It's a double-edged sword.

For any situation to work, there has to be trust, and in this case, we start with two strikes against us, each party uncertain of the other. And what is the result? Look at how the population is laid out in any major city. We have to go back to one of nature's fundamental laws. Things that are the same are drawn to each other for more than one reason. Not only do they have a sense of being back in their homeland by being surrounded by people who look and talk the same as they do, but they also have a sense of security.

In nature, one of the main reasons animals form herds is self-preservation. It may sound morbid, but the more of you there are to choose from, the less chance you are of being some predator's lunch. So what is causing this mass migration of humans from their home countries to go and live in a land different from what they are accustomed to? We know they don't want to because the first thing they do when they arrive in a strange country is to seek out others from their home country who have arrived before them and settle in the same area as they have.

-2-

Back to nature's law. Things that are familiar to each other congregate together. Now you are asking yourself, "Well, they were with their own kind back in their home country. Why did they want to leave?" We have discovered the common denominator in this ongoing problem: overpopulation.

This is one of the fundamental laws of nature – do not overpopulate. It is so vital that many species have built-in biological systems that kick in when conditions are not suitable to support any more of any particular life form. Overpopulation can be directly connected to almost every problem facing the planet today.

CHAPTER TWELVE

There is absolutely no doubt in my mind that the rapidly increasing human population will be responsible for the end of life as we know it on Earth today. The world's major powers are aware of this situation, so they are scrambling to find another planet to sustain human life. They realize it is just a matter of time until our world is so depleted that it can no longer support our existence.

As countries worldwide reach the saturation point and people leave them in record numbers, it burdens the countries that are taking them in. Finding them housing, jobs, and medical care are just a few of the problems that arise from this situation, and this is just the beginning. The native citizens now blame the newcomers for taking their homes, jobs, and doctors—not a good situation.

Nature does not have this problem. Canada is just one country, but there are dozens of different species of birds and animals, all living in the same land and continuing to exist as they have for thousands of years. How do they do it? Because they have the same awareness imprinted in their genetic makeup as we do. Their ancestors learned to avoid creatures different from themselves to survive. Nature's law of survival: the smart ones live to reproduce and carry on the species; the stupid ones are something's lunch.

-2-

Everything would be so much simpler for humans today if we would stop trying to go against the laws laid out for us by nature. This is going to be extremely important if we are to survive.

For this to happen, a lot of things have got to change. Unless we get our runaway population under control, more and more countries are going to reach the saturation point, and immigration will increase at an alarming rate. Soon, every country that allows immigration will look like a box of Smarties, every color in the book. Walk down any large city's main street, and you will see a cross-section of the world's population, all doing what they must to survive.

Every day, these people interact with each other and try to accept each other's differences. This is where nature has the advantage over us. Humans have emotions, and we have all experienced anger, fear, frustration, and doubt about situations that arise in everyday encounters. If it involves a person we already feel is hurting our way of life, it can develop into deep resentment. Nature has resolved this problem, so hundreds of birds and animals can reside in the same country without conflict. There is no interaction between them. This eliminates any chance of disagreement breaking out between them. Humans need to do the same. We want to do the same thing as the first act of immigrants when arriving in a new country is to locate other people of their race and settle in the area they occupy.

CHAPTER THIRTEEN

With more and more people arriving in your country daily, you are not only receiving a representation of all the countries in the world. You're getting a sample of all the different personalities, and rest assured, not all of them help take up a collection in church on Sunday. Life is indeed like a box of chocolates. You never know what you're going to get.

Unfortunately, for many individuals displaced to a strange country, the law of survival has played a significant role in their existence. Many had to do whatever was necessary to survive in their home country. If they bring these old habits with them, it will lead to nothing but trouble in their new home. Unfortunately for the latest arrivals, a group of unscrupulous individuals will be waiting for them, looking for someone down on their luck and strapped for cash. Anyone willing will be offered a quick reward for performing some criminal act. This could be robbing a gas station or stealing a car, so the instigator of it all doesn't have to get involved.

As immigration increases, so does the rate of gun violence and car theft in the country. People listening to the news and reading the papers are seeing the names and nationalities of the persons being caught and charged for these crimes and begin to form an opinion about immigrants. The opinion is not a good one.

-2-

It's an endless chain of events; we must break the cycle. To do this, we must realize the one common denominator in this turmoil – overpopulation. The number one fundamental law of nature is the law of survival. Have we become so arrogant that we assume we are the most intelligent beings in the cosmos?

All the information we have acquired has been taught to us by another human, read from a book written by another human or obtained off the internet created by another individual. The earliest evidence of our existence is approximately 300,000 years, so this is the time we have accumulated knowledge. The universe has been evolving for approximately thirty billion years. Given that amount of time, the odds are pretty good that its law of survival is relatively correct.

Overpopulation is one of the significant threats to the survival of any species, and immigration is one of the symptoms. A high population density breeds disease, poverty, and crime in a country. It forces its citizens to look elsewhere for better living conditions and to leave their home country to live in a strange land.

We may not want to admit it, but we are an essential part of nature, and as such, we are influenced by its laws. Every living thing has its territory that it will defend vigorously.

-3-

Ever wonder why you got stung by a wasp for no apparent reason? You invaded its territory! Humans are no different. Whether it is our apartment in the city or our home in the country, we will take steps to protect it. Most things in nature take an aggressive approach to invaders of their space, but humans tend to be more reserved. We prefer to use signs such as 'Private Property,' 'No Trespassing,' etc., rather than personal confrontation.

But we will defend our personal area. This has nothing to do with who you are, where you came from, or your skin color. It is a throwback to nature's law of survival. We are afraid of the unknown.

CHAPTER FOURTEEN

Overpopulation is, without a doubt, the most serious threat any species can face and is directly responsible for most of the other problems facing the planet today. Climate change, air, water, and land pollution directly relate to overpopulation. And let's not forget the situation created when people have to move to a strange country because of unfavorable living conditions—so-called racism.

This could all be avoided if we concentrated on controlling our population. An old saying goes, "If things don't change, they will stay the way they are." That is not the case in this crisis. If things don't change, things will get worse very rapidly. Many countries have already reached the saturation point. Take, for example, the continent of Africa. Many countries have already exceeded their ability to care for their population, which is still growing at an alarming rate.

In my home country of Canada, things are changing rapidly. Although we are fortunate to live in a vast country with large areas of vacant land, the large cities where all the industry and jobs are increasing in size at an incredible rate. Much of this is directly linked to immigration. The City of Toronto's population is growing so rapidly that its infrastructure can no longer handle the traffic overload. So much so that the province's Premier suggests that we will have to build tunnels under our existing highways to handle the increasing traffic volume. This would not only take years to complete. It would be an

-2-

Additional burden of billions of dollars on the country's citizens in taxes. And this is just the beginning.

Immigration is a process that requires extensive forethought and planning before being initiated. What the new arrivals will need must be in place before they set foot in their new country. Immigrants sleeping on the streets and living in tents in local parks do not sit well with the local population and do nothing to improve the opinion of new citizens.

It is proven that all species of life will go to great lengths to be with their own kind. So what happens to a country when it is populated with races of people from all over the world? Think of what it would be like if your relatives came for the weekend and decided to stay. These are people with the same beliefs as you and are part of your family, but after a while, it gets to the point that you want them to leave. They are in your space. This goes back to the fact that every living thing has its own territory. Let's imagine for a moment that these people have different beliefs than you, and look and talk differently than you and come into your territory to stay. I am confident that you would be highly agitated, and again, this has nothing to do with racism. Remember when you were a child and going outside to play, and your parents would warn you not to talk to strangers? This is the same situation. It's nature's law of survival kicking in. Beware the unknown.

-3-

After thousands of years of evolution, almost everything we do is controlled by nature. No matter what life form it is, once it reproduces and its offspring is strong enough, it packs its bags and goes on its merry way. It's the same with humans. When our children are old enough, they go off on their own. This is nature's way of preserving the species. A smaller group requires fewer things essential for life than a large one. If disaster should hit, it would only affect the people in that vicinity, not everyone. Also, the environment in which they live will be able to sustain a small group for a more extended period than a large one.

The oversized item on the agenda of every level of government right now is the staggering increase in violence in every densely populated section of the world. In their haste to find the cause of this behavior, they have come up with what is behind it all. Guns. In their ignorance, they have even gone so far as to label this activity 'gun violence.' Brilliant.

I see this whole situation from a different perspective. My family grew up off-grid. We didn't even have a grid to connect to. We had no hydro, phone, plumbing, or modern convenience. You have never experienced pure joy until you have planted your butt cheeks on a frosty hardwood plank with a round hole cut in it at forty below Fahrenheit. That is true bliss!!!

-4-

Everything on the homestead had a purpose, and guns were no different. They were simply a tool with a job, the same as an axe or a saw. They were used as a deterrent to keep wildlife out of our vegetable garden, protect our livestock, and put food on the table. There was always a rifle hanging on the wall over the kitchen table. I still legally have guns in my house to this day, and not once in all my years has a gun ever tried to harm me.

Gun violence – no!!! It's people's violence that is driven by frustration and desperation from being packed together like sardines in strange cities that are different from them and with no idea of what is in store for them in the future. It's easy to target guns as the culprit. Handguns are small, light, easy to conceal, and easy to obtain on the streets. Unfortunately, they are not the root of the problem. They are just the tool used by humans to commit crimes.

CHAPTER FIFTEEN

The governments of today unanimously agree that the solution to crime in our countries is gun control. Nothing could be further from the truth. There will always be a certain percentage of the population that, for some reason, will want to make a living by committing a crime. Control this segment of the human race, and the guns will all behave.

So what does crime, violence, war, starvation, climate change, and so-called racism have in common? Overpopulation. Pack too many species together, and you will have a problem. The earth has a saturation point, and we are quickly approaching it.

Unfortunately, no matter how large, the planet cannot keep supplying us with what we need to exist, especially when the demand keeps growing. The current trend is for the overflow from the countries that have reached their saturation point to migrate to the parts of the world where there is still some room available and that are willing to accept them. This is a solution that is doomed for failure.

Unfortunately, the human race has an obsession with sex. I don't have any statistics on this activity, but I think I can safely say that it is the most practiced physical activity in the world today. Sex was meant to be enjoyable. If engaging in it felt like hitting your thumb with a hammer, the species would go extinct, but the fact is that humans are the only species that are so possessed with performing the act.

-2-

Fornicating was not intended to be a pastime but a necessary procedure to ensure a species would not go extinct. Many species only participate in the act of reproduction once a year. Right now, you are celebrating that you are not a moose!!!

All joking aside, mating in nature is a severe business for many species, with the males competing violently for the right to mate with as many females as possible. This is not for pleasure but to ensure the survival of the species.

Humans don't have to worry about replenishing the population. This is why immigration is only a temporary fix. When people leave a country, the remaining inhabitants get busy replacing them.

So, how do we stand as a species at this point? Many countries have reached the saturation point population-wise, and the citizens are lining up to go to any country that is still willing to accept them. We have a human population over billions of years of evolution programmed biologically and genetically to distrust beings that are different from them.

-3-

How can this possibly work? It can. It took millions of years, but nature figured it out, and we have the blueprint right before us. First, we must learn how to control our exploding population. We are experts at controlling the numbers of other life forms, so much so that one million of the eight million species on the planet are on the verge of extinction, and hundreds more are already extinct. We are very good at controlling things. We control our pet population by getting them spayed or neutered. Farmers control their livestock numbers to afford to feed and house them. And we love to control people. That's why governments pass so many laws.

So why can't we control our own population? The answer is straightforward. Just about everything we see, read, watch on TV, or buy is based on sex. And the reason is simply that sex sells. From attractive actors on TV and in movies to scantily clad performers in live shows, and even the very clothes that you wear, everything is geared to stimulate our hormones.

But this alone is not the only reason for our exploding population. Even today, governments reward us for having children by issuing a monthly check for each child in the family. Have another child, and get another check. Also, each child saves you money at tax time as an additional tax deduction. Sweet!

CHAPTER SIXTEEN

Probably the number one reason for our rapidly growing population is the lack of education in birth control and the unavailability of birth control products in many countries in the world. Mix this with the fact that in the Hindu religion, the woman must have children, and if the men have a vasectomy, they are considered inferior. Add this to the fact that there are 1.2 billion Hindus in the world, and that amounts to a lot of babies.

Many of the world's religions do not promote birth control. The Islamic and Buddhist faiths believe in having large families, and until very recently, the Catholic church believed that the only reason for sex was for reproduction, so reducing the population has some hurdles to clear.

Many life forms in nature have an advantage over humans regarding population control. They use the predator/prey equation to create a cyclic system to control their numbers, which works exceptionally well. I will give you one example. In many northern regions, there is a species known as the varying hare. It was called this because its color varies between brown in spring and summer to match its surroundings to white in winter to match the snow. This has been an attempt to camouflage itself from its predators. It is also known as a snowshoe hare for its large furry hind feet, allowing it to travel over deep snow easily. Its life cycle ranges from times of low numbers to periods of abundance.

-2-

These numbers also affect the population of the coyote, fox, and wolf, the hare's main predators. If the hares' numbers are low, predator numbers dwindle, and bunnies return. With the return of their food supply, the predators rebound, and the cycle starts all over again. This scenario occurs millions of times daily, from the bottom of the oceans to the plains of Africa and the jungles of South America. Nature loves a balance, which is one way of achieving it to keep numbers in check.

It has been a long time since we were running for our lives, being chased by a saber-toothed tiger and getting a grip on what it feels like to be the prey, but it hasn't stopped us from murdering thousands of our own species every day. Why is this happening? Because we're all being thrown together like a can of mixed nuts. We are all homosapiens, but we are all different flavors. Have we not learned anything from nature? Birds of a feather. HELLO!!!

CHAPTER SEVENTEEN

It seems that somewhere along the evolutionary road, humans have forgotten the number one law of the universe. The law of survival. This law was enforced very early with a straightforward rule. Stay away from things that are unfamiliar to you.

Well, it seems we have had our reawakening of rule number one! The government of Canada's Immigration Minister has announced that his department made a significant miscalculation on the number of immigrants allowed into the country and dramatically slashed the numbers for the next few years.

As I write this, a new administration is in the process of taking over the United States and is promising a massive deportation of immigrants from that country. Why the sudden policy change? Most individuals want to help another human being, but when food prices and interest rates are high, and there is a housing shortage, people start to point fingers. We have people sleeping on the streets now. Why are we bringing in more immigrants?

This would never happen in nature. It controls its population to reduce the need for a species to co-exist in another life form's territory. Certain things in the universe are constant, such as the speed of light in a vacuum and the charge of an electron.

-2-

Since the very beginning, when life forms began to roam the earth, all species remained with their own kind to survive. I believe this is another constant of the universe. I'm not exactly sure how long something must remain in an unchanging condition to qualify as constant, but I feel that a few million years should be eligible. I think thousands of species on the planet would agree. So, if the preferred arrangement is for like-types of a species to live with their own kind, why do humans try to break the number one rule of survival?

This returns us to the common denominator that started all the problems: overpopulation. When the surge of immigrants started from countries whose populations had reached saturation, other countries were happy to take the overflow. Canada was one of these countries. They wanted a diverse population to serve the rest of the world and that Canada was a warm and inviting place. This train of thought is slowly being reversed.

CHAPTER EIGHTEEN

It takes many things to survive on this planet, and not the least is a country to survive in your country. A place you call home, a sanctuary where you feel safe and needed, go shopping without worrying about being shot, go to bed at night, and feel secure.

We used to have that. Flashback a few hundred years, and the indigenous people of North America found themselves in the same situation as we are facing now. Strange humans were landing on their shores.

I previously touched on the following events, but some pertinent information has compelled me to revisit them more deeply.

As the native people lived very close to nature, the law of survival took over, and the new threat made them uneasy. The Portuguese arrived first in the fifteenth century but found the conditions too hard and left. The Irish, Scots, English, and French soon followed them. The peaceful country of the indigenous people was starting to slip away. It didn't take them very long to realize that their worst fears were about to become a reality. This was no one-day guided tour to see the natives. They intended to stay, and stay they did. Not only did they stay. They started a complete takeover.

-2-

The newcomers encroached on the natives' best hunting and fishing areas and displaced them from the best places to have their teepees and villages. Initially, the invaders acted like they wanted to help the natives and tried to befriend them. But this was all an act to earn their trust to set up a trade for their beautiful pelts and hides. This worked for a while, but as the white man continued to arrive and push deeper and deeper into the new land, the resident population soon realized that they had no intention of leaving. They intended to stay.

Hundreds of skirmishes followed, but it was a no-win situation for the natives as they were severely outnumbered, outgunned, and shortly pushed further back. We all know how this story ended, which wasn't good for the home team. They ended up losing their country and being regulated to specific areas of land known as "reserves" for them to exist.

Racism had absolutely nothing to do with the conflict, as mentioned earlier. Before it started, neither side knew the other one existed. I say it simply for two reasons. The law of survival was correct, making the natives very uneasy when they first saw the strangers. And number two, to show what can happen when an invasive species enters a new environment. To have ill feelings toward someone, you must have some information about that person or person. It is incidents like this in history that influence the way we think.

CHAPTER NINETEEN

When you get right down to the bare-bone facts, racism has very little to do with a lot of the conflict in the world today. One perfect example is the ongoing war between Russia and Ukraine. Russians do not hate the Ukrainians. They need their land. Russia's population is growing, but agriculturally, it cannot sustain its population.

On the other hand, Ukraine is blessed with rich farmland that they are not about to give up. The Ukrainians have it; Russia wants it. The bully will try to take it. Classic schoolyard tactics.

The same can be said for the numerous conflicts occurring in and between small countries throughout Africa and South America. The people in these countries live in very unsatisfactory conditions and the leaders of these countries and whoever supports them are held responsible for the mess. In these situations, it is neighbor pitted against neighbor and it is simply an act of desperation. There is no racism here.

In the 1500s to 1800s, an act of absolute disgust was committed towards the black population of the world. We called it the "Slave Trade." Hundreds of thousands of black men, women, and children were collected from Europe and Africa and shipped to America to be sold to wealthy white people. The death toll in this catastrophe was staggering. The living conditions on the ships with the storms, rats, and disease were unbearable, and thousands perished to be buried at sea. Probably a blessing.

-2-

The ones who survived the crossing were auctioned off to the highest bidder to work as enslaved people. But this wasn't the worse of it because they were now considered low-life and not allowed to live as white people. They were not allowed to perform most normal day activities. For example, they were not allowed to leave their owner's property unless accompanied by a white person. Most degrading, they were not permitted to learn to read or write. A relationship with any white person was forbidden. Black marriages were not considered legal, and blacks were not allowed to buy or sell anything. Owning a weapon of any kind was almost always a death sentence. Disobeying any of these rules guaranteed you, at the very least, a brutal flogging. This is racism at its very worst.

Over the last fifty years, the makeup of classrooms in our education system has changed dramatically due to the massive influx of immigrants. I grew up in a small farming community where all the students were of the same nationality and still had disagreements. And there always will be. Now, with the classrooms filled with a mixture of races, colors, and beliefs, if the parents of a student are uncomfortable with the parents of another student in their child's class, these feelings will likely be passed on to the uneasy parents' child by the other student and may cause conflict.

Once again, the law of survival is warning us to be aware.

-3-

Another situation that is fresh in our minds is the conflict between Israel and the Palestinians. Israel claims it is aiming its attack against Hamas and not the Palestinians, but thousands of them have been killed. Let's dig a little deeper. It states in the Bible that Abraham promised land to the Israelites that would be their forever home, while the Palestinians were left to wander homeless. This decision has caused tension between the two sides ever since. Now, the killing of thousands of innocent Palestinians in Israel's war against Hamas is fueling hatred against Israel around the world. This has nothing to do with nature's law of survival. This is not racism. This is revenge against Hamas on Israel's part and anger and hate for Israel by the Palestinians for what is happening to them.

CHAPTER TWENTY

Probably, our best hope of putting an end to racism is by understanding what it is in the first place. Racism is the act of causing mental or physical suffering to any person or persons because of their race, color, or beliefs. No one is born a racist. It is a quality taught to us by our environment.

Everything we read, see in movies, on TV, hear on the radio, or see on the internet influences how we feel toward others. Also, how other people have treated us in the past is a significant factor in our opinion of them. No one is born a racist. We are born with an evolutionary will to survive and a mandatory need to avoid a different life that could cause us harm.

This is the number one rule of nature, the law to survive. This is the job of every living thing on the planet. To keep taking one more breath. This is why there is no racism in nature. Everything is too focused on staying alive to be concerned about what the rest of nature is occupied with. The blueprint laid out by nature is there for us to follow, but we are losing our grasp of where we came from. It seems we've become too involved in what everybody else is doing that we forget to take care of number one and concentrate on keeping a safe distance from unfamiliar things.

-2-

What is fascinating to me is that by using modern technology, "lidar," archaeologists are uncovering magnificent cities built by advanced civilizations deep in the rain forests of South America hundreds of years ago. This was before the Inca's and Maya's reign in South America and they amazingly vanished. This is happening all over the world. One archaeological dig in Turkey called Gobekli Tepi is approximately twelve thousand years old and in the stone age era and is revealing structures that rival our ability today. Where did the civilization that constructed these places go? They certainly didn't just keep improving, or they would still be around today. There was little chance of conflict being responsible for their demise as they had no competition, so racism was not involved.

The one constant in all this is that all that disappeared was human life, and all other forms of nature survived. There definitely must be a problem in our programming. If nature can continue after billions of years and we keep disappearing after a few thousand, maybe we should rethink what we are doing, back to the drawing board.

CHAPTER TWENTY-ONE

Initially, we started following nature's laws with similar humans staying with their kind, but somewhere along the road, we gave in to temptation. One group would take a liking to another tribe's territory and try to take it for their own, and often would succeed. This was the beginning of the mixing of different types of humans.

This is still happening today, only on a larger scale, countries trying to take over other countries. After Columbus discovered America, the flood gates were opened. The lure of land free for the taking was just too strong and the immigration to the new world was underway. What happened next was absolutely appalling. The indigenous tribes didn't have a chance against the onslaught of newcomers. Any that tried to resist the takeover of their land were killed. The ones that didn't eventually succumbed to the multitude of diseases that the intruders brought with them. As they had never been exposed to any of these new maladies, they had no immune system and their bodies had no protection against them. They were helpless. Whether this was happening in what is now Canada, the USA or South America, the end result was the same – total devastation for the native population. In reality, in the period of time from 1500 to 1800, many of these ancient tribes went extinct.

-2-

Now is it just me or is there an ongoing pattern happening here that has been going on ever since we came into existence. What it is couldn't be more obvious. Whenever we encounter life forms different from ourselves, there is the very distinct possibility that something unfavorable is about to occur. If you don't believe me, ask the caveman hanging out of the mouth of a dinosaur. That dinosaur didn't pick the caveman up because he looked like he was tired of walking. He picked him up so he could have lunch.

From that simple action, we have learned to be wary of things that are different from us and it is our duty as a civilization to be very careful not to generalize.

Chapter Twenty-Two

This is where nature has us at an advantage. It has not as yet started the mixing of different species where humans are doing it more and more. This is extremely hard to understand as we are a part of nature. No species in nature wonders whether it is okay for robins to live with blue jays. It just doesn't happen. Too bad it doesn't work that way for humans.

You can't let one dinosaur into your yard because he's pretty. A cute dinosaur can eat you the same as an ugly one. It works that way for immigrants too. A smiling one can harm you the same as a grumpy one. You can't take a chance on any of them without more information. You have to be very cautious with who you trust.

There is one very important thing to remember. None of these people are coming to your country to improve your way of life. They are coming because conditions in their country have become unbearable and they are looking to move into an environment that will allow them to achieve more than they can in their home country.

There is one more item that we tend to forget, and that is the fact that the people coming to your country have an evolutionary history, the same as you. They have the same misgivings about you as you do about them. It's difficult when one person in a relationship is uncertain of the other, but when both are uneasy, it's next to impossible. Nature has it right. Just keep them apart.

-2-

It is not by coincidence that all species of life on earth choose to exist with their own kind. It took billions of years of evolution to develop a formula for survival that works, but we have the proof right in front of us. At least for now.

Humans are extremely good at taking something that works perfectly well and making it less useful. Statistics show that wild life numbers in the last hundred years have declined by fifty percent. Why? The explosion of the human population.

With more and more people every year, there is a decrease in the amount of habitat that there is for wild life. Not only is habitat being reduced, but whole areas that are suitable for nature are being eliminated for human use. We only have one planet. What makes us think that it's here just for us to do whatever we want, and too bad about everything else? If we keep going against the laws of nature, it's going to come back to bite you. The law of survival is simple. Stay away from things that are different from you. They may not harm you, but they could! Why take the risk? Staying away increases your chance of survival.

Why do we as humans try to force all races of people to live together when all other life forms on earth stay with their own kind? We already have a two-way evolutionary mistrust between us. It's a no win situation.

-3-

Between forcing races of people to live together that already have a mistrust of each other to blindly causing the extinction of dozens of species of wild life, it is obvious that we are ignoring all the rules. We are out of control. The conditions that exist in the countries that are forcing people to leave only change one thing. Their country of residence. All of their beliefs, religions, inhibitions and hates remain the same. Any conflicts they had or feelings they felt towards other people have not changed The only difference is that they now take place in another country.

At the present time, the whole civilization is just living in the present with no thought to what is inevitable. We are like a rapidly expanding balloon. The earth will reach its saturation point and we will go the way of all the ancient civilizations before us. The emotion that we call racism is a symptom of the overcrowded planet crying out for attention. Every species on this earth has to control its population. Nature plays a big role in this with predators, disease and climate change, but humans play a larger role, controlling to the extent that many species have gone extinct.

In this case, it is not control. It is elimination. Unless we get our act together, we will self-destruct unless some cosmic event does it for us one more time.

-4-

What are the consequences of overpopulation? The more humans on the planet, the faster the earth's resources will be depleted. More contaminants will go into our lakes and rivers. Our atmosphere will become saturated more quickly with greenhouse gases and there will be an ever increasing demand for electrical power, forcing us to use nuclear generation plants. And how do we get rid of the nuclear waste? We hide it in the earth, making it unfit for anything to live on or in.

Also with more and more people, our habitable areas of the planet are disappearing at an alarming rate. With all of these negative events taking place, you would think that we would catch on to the fact that something is rotten in Denmark. But we don't! We just keep on making humans! With more people being displaced from their home land by over population, poverty and unrest and going to live in a strange country with a mixture of cultures from all over the world, conflicts are sure to occur. This is not racism. It is frustration.

CHAPTER TWENTY-THREE

If there's anything good that can be said about racism, it is the fact that it is not directed at an individual but rather at their skin color or the beliefs that they have that are different from yours. As I have mentioned earlier, no-one is born a racist. This is feelings and emotions passed down to us from our ancestors.

A perfect example of this is the hate that is being directed at Jewish people all over the world. This has been fueled by the war that Israel is waging against Hamas and the Palestinian people living in Gaza are suffering terrible losses because of it.

But this is by no means the genesis of this problem. Hatred against Jews has been going on for hundreds of years. Probably the best example of this was the Holocaust in which Hitler tried to eliminate the Jewish race and thousands of Jews were executed.

Religion has probably been responsible for more conflict over the course of history than any other single factor. Punishing people for their choice of faith has been handed down through the generations and whether you decide to call it 'religious discrimination' or racism is your choice.

-2-

At the present time in China, this scenario is being played out against a race of people who call themselves the Uyghurs. China is a very densely populated country that practices many forms of religion, including a large portion of the population that are atheist. The approximately twelve million Uyghurs (pronounced Wayhurs) are Islamic. The experts say they are being persecuted because of their faith in Allah. So how come none of the other different faiths in the population are being targeted? I believe it is something else.

The Uyghurs are unique. Their appearance is different from the Chinese and they even have their own language. One more thing, they keep to themselves. Is any of this starting to sound familiar? It is nature's law of survival kicking in once again. The Uyghurs are different and are trying to stay away from the Chinese population, but the country is so over-populated that there is no place for them to escape to. As for the Chinese, nature's law states we should avoid things that are not the same as us, but in this case, that is impossible. The only other alternative is to make life so unbearable for the Uyghurs that they will eventually find another home and move on. In every case of racial discrimination, if you research deeply enough, you will find that our evolution and our will to survive is playing a major role.

Take a close look at the slave trade. Not only did the Negro population differ from us in color, but in most cases, in dress, language and culture as well, and this made us uneasy to the point of being fearful of them. We couldn't get rid of them so the next best thing was to control them.

-3-

This was our way of convincing ourselves that if they served us and had no rights, they would be incapable of causing us any harm—the law of survival.

Even in the most disgusting act of genocide the world has ever known (the Holocaust), there is a connection to the law of survival. In his early days before he came into power, Hitler worked as a painter with a group of people who hated the Jewish population. After listening to all their horror stories about the terrible acts committed by the Jews and how they were going to take over the earth, he became convinced that they were a threat to him and his country.

When he eventually came into power, he set out to do something about it by trying to eliminate all the Jews on the planet and therefore saving himself and the German population. It may have been the act of a mad man but he was driven by the will to survive. The survival instinct doesn't dictate that you should eliminate all living things that are different from you to protect yourself from harm. It is only an uneasy feeling that tells you that this is something to stay away from.

In Hitler's case, he was just a mentally deranged person who wanted to rule the world. As technology advances farther and farther, it is very easy to forget where we came from. It's hard to believe there was a time when our only job was to survive until the next day. The law of survival was extremely important then and it still is today in nature.

-4-

We have to take a moment and stop and reflect on the fact that the first million years or so of our existence were spent running for our lives to avoid becoming something's lunch. Once we do, it's easier for us to realize that our rejecting someone is not racism. It's nature telling us to proceed with caution.

After scrutinizing all the information and concentrating only on the relevant facts, here is what we are left with.

After millions of years of evolution, we have arrived at this point in history with an embedded fear of the unknown and a preference to avoid life that is different in appearance from us. This is the result of unfortunate incidents that have happened to us in the past.

CHAPTER TWENTY-FOUR

Unfortunately, many of the old world countries that have been in existence much longer than the countries that were established later (USA and Canada) have reached their saturation point as far as population goes. Their lands have become depleted and people are living in poverty. This is forcing them to seek refuge in any country willing to take them, looking for a better quality of life.

This is not always the case. Check out any large city in any country with an immigration policy and you will find people sleeping on the streets. This is because of poor management and allowing more people into the country before it is prepared for them. What a great way to meet a person that is different from you, by having them live on the street in front of your home.

I can't think of anything that enhances the feeling of affection more for an immigrant from a different country than for the person to show up unannounced and live in a tent on your lawn. All kidding aside, as I've said many times, no-one is born a racist.

What happens is, we are influenced by stories by the media, shows on TV, what we read and events that have happened in the past that cause us to form negative opinions of others. Basically we have two options. We can carry on the way we're going with no certain outcome or go back to our roots and take the path taught to us by evolution. Life forms that are the same remain together. Everything in nature can't be wrong.

-2-

If your Mexican neighbor's Chihuahua is using your front lawn as a toilet, letting your grass grow in the hopes that the dog will get lost in there and starve to death is not racism. It's saving the environment from greenhouse gases by not running your lawn mower.

www.ingramcontent.com/pod-product-compliance
Lightning Source LLC
Chambersburg PA
CBHW071517210326
41597CB00018B/2790